# MATH BRAIN TEASERS

## An Illustrated Collection of Original Puzzles

Copyright © 2021 by Unicorn Books

All rights reserved. No part of this publication may be reproduced, stored or transmitted in any form or by any means, electronic, mechanical, photocopying, recording, scanning, or otherwise without written permission from the publisher.
It is illegal to copy this book, post it to a website, or distribute it by any other means without permission.

First Edition

# Table of Contents

Introduction..............................1

Appetizers for the Mind.........3

Intense Brain Workout.........45

Dangerous Territory..............79

Your Free Gift........................109

# Introduction

This book consists of more than a hundred pages of brain teasers. But it is not simply a list of problems. Each high-quality puzzle is accompanied by meaningful illustrations, useful hints to guide your reasoning, and a complete explanation of the solution.

The puzzles are organized into three sections, corresponding to increasing levels of difficulty. But it is not necessary to follow a strict order. Flip through the pages and let yourself be inspired by the titles and images in choosing your next challenge.

This book was designed to be at the same time challenging, entertaining, and fun. The riddles don't require any advanced knowledge and use only elementary math. But don't underestimate them. They can be hard for people of any age and level of instruction. Don't believe me? Try them on your parents, your older siblings, or even your teachers.

Discussing a puzzle with others is certainly the most efficient and enjoyable way to get to the solution. For this reason, the book also represents an alternative way to spend time with friends and family, engaging them in a fun and entertaining activity.

You will notice that the more puzzles you tackle, the better you will get at solving them. That's because you are improving your ability to think outside the box, which is useful in every aspect of your life.

Before we start ...

If you wish to support our work and would like to see more books like this one available, please

leave your honest review.

And now... Let the fun begin.

# Part 1

# Appetizers for the Mind

# Toy Zoo

Cooper has an impressive collection of 100 toy animals. One day he arranges them into 10 rows of 10 animals. He chooses the tallest animal in every row, and then picks the shortest among these 10. The winner of this contest is Jumbo the elephant.

Cooper then tries a different procedure. He takes the shortest animal in every column and then picks the tallest among these 10. In this way, he ends up with Bridget the giraffe.

**Which one is taller, Jumbo or Bridget?**

# Hint

First consider the case in which Jumbo and Bridget are on the same row or on the same column.

## Answer

Take the toy that is in the same column as Bridget and the same row as Jumbo. Jumbo is taller than this toy, while Bridget is shorter. We conclude that Jumbo is taller than Bridget.

# Growing Bacteria

Matilde is a scientist working on a newly discovered species of bacteria. These bacteria triple in number (and therefore in volume occupied) every thirty minutes.

At 8 am Matilde puts a single bacterium in a test tube. At 5 pm the test tube is completely filled with bacteria.

At what time would the test tube have been full if Matilde had put three bacteria in the test tube instead of a single one?

# Hint

How are the two scenarios related? At what time were there three bacteria in Matilde's test tube?

## Answer

The test tube would have been full at 4.30 pm. In the first scenario, the test tube contains three bacteria at 8.30 am. Hence the second situation is the same as the first just half an hour ahead.

# Bedtime Story

Every night, grandma reads to Sydney part of a fantasy novel. The book they are currently enjoying is 96 pages long.

Grandma usually goes through at least 10 pages in one night. Tonight, however, Sydney is exhausted, and grandma manages to read only two pages before Sydney falls soundly asleep. The sum of the digits of the numbers of the two pages is 24.

**What pages did grandma read that night?**

# Hint

The first digit of the two numbers can't be the same. In fact, if that was the case, the sum of all four digits would be an odd number. Can you figure out why?

Since the sum is 24, that is, an even number, we only need to check the pairs 19-20, 29-30, 39-40, and so on.

# Weekend in Space

Five spaceships left the planet Gok for a weekend trip to another corner of the galaxy. Unfortunately, on the way back, one collides with an unidentified object, resulting in an engine failure. Its passengers, therefore, transfer to the other vehicles. They do it so that each of the remaining spaceships will have the same total number of passengers.

While abandoning the damaged vehicle, one of the passengers observes, "If there had been one less passenger in this spaceship, then it would have been impossible for us now to distribute in this way".

**Since the four undamaged ships had respectively 17, 19, 14, and 22 passengers, how many people were traveling in the fifth spaceship, at minimum?**

# Hint

In the end, each of the four remaining spaceships has the same number of passengers. What is this number? Remember that it is the smallest possible.

## Answer

Each spaceship returns to Gok with 22 passengers. Since 22-17=5, 22-19=3, and 22-14=8, the fifth spaceship had 5+3+8=16 passengers.

# New Releases

Betty and Susan are discussing the latest novel of their favorite author. It was released just last week.

**How many pages does the book have?**

# Hint

It might be helpful to write down some multiples of 17.

## Answer

The book has 136 pages. Betty finished it in 8 days (17×8=136), while Susan finished it in 12 days (11×12+4=136).

# French Cuisine

Damien is a famous French chef. He has just received a mouth-watering wheel of cheese in the shape of a perfect cylinder.

Damien is ready to dip his knife into the cheese. He has a precise task to accomplish, which is to obtain eight identical parts. In order not to lose his chef license, he must follow strict rules when using his knife. He can only perform straight cuts, and he is not allowed to rearrange the pieces between the cuts.

**What is the minimum number of cuts that Damien has to make?**

# Hint

The cuts need not be "vertical".

## Answer

Three cuts are enough. Damien makes two vertical cuts along two diameters to get four identical pieces. He then slices the wheel horizontally to obtain eight pieces.

# Crowded Building

Eric lives with his dog Pebble in apartment number 46, on the fourth floor.

In his building, apartments are numbered starting from the first floor. Moreover, each of the five floors of the building has the same number of apartments.

**How many apartments are in Eric's building, at most?**

# Hint

To maximize the number of apartments on each floor, Eric must live on the apartment with the lowest number on his floor.

## Answer

For Eric's apartment to be the first of the fourth floor, there must be 15 apartments on each floor. Therefore, there are 15×5=75 apartments in the entire building.

# Street Signs

Benjamin is in the car with his family. They are driving toward their house in Seymour. On the way there, they will stop to visit Benjamin's grandparents in Norwood. At a certain point, Benjamin sees the following street sign.

After a while he sees another similar sign giving the distances to Norwood and to Seymour. He notices that the two numbers are now made of the same two digits. The only difference is that in one of the two there is a point between the digits.

**What distances are indicated on the second street sign?**

# Hint

What will a similar sign show after they have traveled one more mile? And after two miles?

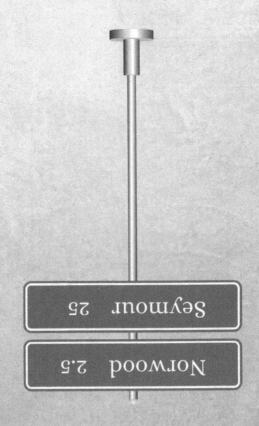

## Answer

The sign will appear after six miles and will be as follows

# Sweet Exchanges

Chloe loves gummy bears, but not the red ones. Luckily, those are the favorites of her friend Emma. The two girls agree to exchange on the following terms: for every three red gummy bears that she receives from Chloe, Emma will give back five of different colors.

At the beginning, Chloe had 37 gummy bears. After exchanging with Emma, she has 53, none of them red.

**How many red gummy bears did Chloe have initially?**

# Hint

How many gummy bears does Chloe earn from a single exchange? Can you figure out how many exchanges were made?

## Answer

Chloe has 53-37=16 gummy bears more than she had at the beginning. Since she earns 2 gummy bears in every exchange, there were 8 exchanges in total. Therefore, she had 8x3=24 red gummy bears.

# Colorful Runs

Mason, Samuel, and Gabriel have just raced each other to determine who is the fastest runner. Here is the information we have. Samuel is wearing a green T-shirt. Gabriel didn't win. The one with the blue T-shirt finished in third place. Mason is not wearing a yellow T-shirt.

**Which of the three friends won the race?**

# Hint

You might find it helpful to organize the information in the following grid.

|         | Yellow | Green | Blue | 1 | 2 | 3 |
|---------|--------|-------|------|---|---|---|
| Gabriel |        |       |      |   |   |   |
| Mason   |        |       |      |   |   |   |
| Samuel  |        |       |      |   |   |   |
| 1       |        |       |      |   |   |   |
| 2       |        |       |      |   |   |   |
| 3       |        |       |      |   |   |   |

## Answer

Samuel won the race.

# Fast Sales

Today was a profitable day for Gloria. She got to the market with her load of watermelons early in the morning, but after two hours she had already sold everything.

The first customer bought half of her watermelons. The second one bought one third of the remaining fruits. The third customer asked for five watermelons, but Gloria gave him an additional one for free, so that she had no watermelons left and could go home.

**How many watermelons did Gloria have at the beginning of the day?**

# Hint

Reason backward. Gloria had six watermelons before the last customer came. How many did she have before the second customer?

## Answer

Before the second customer bought one third of the watermelons, Gloria had nine. Since the first customer bought half of her initial number of watermelons, she had 18 at the beginning.

# Free Food

Amy, Donna, Kimberly, Matthew, and Tim all love pizza. When they found out that the newest pizza place in town would give out yearly coupons to their first customers, they decided to show up there the night before. Now the five friends are standing in line in front of the doors of the pizzeria.

Amy and Kimberly are not next to each other. Donna is not next to Amy, Kimberly, or Tim. Tim has a spot before Matthew, but he is after Kimberly.

Unfortunately, the restaurant will give out only two coupons, so only the first two in the line will receive one.

**Who will get the coupons?**

# Hint

Start from Donna. Given that she is not next to Amy, Kimberly, or Tim, she must be either the first or the last of the line.

## Answer

Donna is next to Matthew, and they can't be in first and second position since Tim has a spot before Matthew. Therefore, Donna and Matthew are last in the line. The order of the five friends in the line is the following: Kimberly, Tim, Amy, Matthew, and Donna. Kimberly and Tim will receive a coupon.

# Efficient Chains

Mark the blacksmith has four pieces of chain, each consisting of three rings. He wants to obtain a single round chain as in the picture.

It takes Mark a minute to open a ring, and another minute to close it.

**How long will it take Mark to complete this task, at minimum?**

# Hint

Mark doesn't need to open rings on all four pieces.

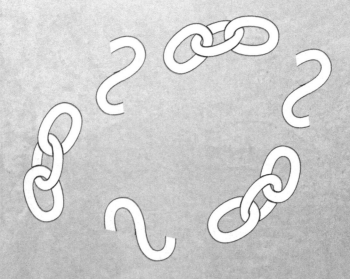

## Answer

It will take Mark 6 minutes to make the chain. He opens all three rings in one piece and uses them to connect the remaining three pieces together.

# Stolen Berries

Kayla walks home after an afternoon spent looking for wild berries. She is quite happy because she has improved on the 16 berries found yesterday. However, she still hasn't been able to beat her brother's record of 40. Once home, Kayla leaves her basket containing strawberries and blueberries on the doorstep and goes inside to say hi to her mom.

When she returns outside to get her basket, she finds an unpleasant surprise waiting for her. A bird has eaten 2 out of every 7 of her berries! Now, she notices, she has as many strawberries as she has blueberries.

**How many berries were in Kayla's basket when she got home?**

# Hint

Consider the sentence "A bird has eaten 2 out of every 7 of her berries". For it to make sense, the total number of berries must be a multiple of 7.

## Answer

The only multiples of 7 between 16 and 40 are 21, 28, and 35. After the bird has eaten 2 out of every 7, the number of remaining berries in each case would be 15, 20, and 25, respectively. Since there must be an even number of berries, 20 is the correct alternative. Therefore, when she got home, Kayla had 28 berries in her basket.

# Cheerful Birds

The big cherry tree in my garden is home to many birds. This morning I observed several Robins and Cardinals on its branches.

Suddenly, my dog started barking. Eight Robins got scared and flew away. Five of them later came back. When my dad started his car, three Cardinals went away. After a few minutes, two came back to the tree.

At that point, I counted a total of 13 birds on the tree.

**How many Robins were on the tree initially, at most?**

# Hint

How many birds went away without coming back to the tree? How many birds were on the tree initially?

## Answer

At the beginning there were 17 birds (the 13 that are there now plus the 4 that flew away without coming back). Since we know that at least 3 of them were Cardinals, the maximum possible number of Robins is 14.

# Curious Birthdays

In a group of friends, everyone has a different birthday.

However, summing the two numbers corresponding to the day and month of birth of each friend, one always obtains 35.

**How many friends are in that group, at most?**

# Hint

There aren't many possibilities for dates where the sum of the month and the day gives 35. Try to find them all.

## Answer

The possible birthdays are: 5/30, 6/29, 7/28, 8/27, 9/26, 10/25, 11/24, 12/23. Therefore, there are at most 8 friends in the group.

# Wise Substitutions

Last Sunday my favorite soccer team played the championship final. They started the game with the players numbered from 1 to 11. Sitting on the bench were players 12 to 18. At halftime, the coach decided to replace three of the players in the field with three of those sitting on the bench. The sum of the numbers of the players who went out is 21.

I noticed that summing in pairs the numbers of the players who went out I could obtain the numbers of the three players who entered the game. They ended up winning, but I was a bit sad that my favorite player, who had number 13, never left the bench.

**What are the numbers of the players who entered at halftime?**

# Hint

Players number 10 and 11 weren't involved in the substitutions. In fact, if that was the case, the sum of the numbers of the other two players who were replaced would be 21-10=11 or 21-11=10, respectively. That is absurd, because that sum should correspond to a player who was on the bench.

## Answer

The players who entered at halftime had the numbers 12, 14, and 16. The players who went out had the numbers 5, 7, and 9.

# Unique Clock

Blake's clock is very peculiar. While the hour hand and the minute hand work normally, the second hand runs counterclockwise. Blake looks at his clock at 11.30 am. In that instant, it shows the right time.

Blake keeps staring at the clock until precisely 12 pm, when it again gives the correct time.

**How many times during that half hour (including the initial and the final moment) was the clock showing the correct time?**

# Hint

How many times in a minute does the clock show the correct time?

## Answer

Blake's clock shows the correct time whenever the second hand hits the 12 or the 6. In 30 minutes, this happens 61 times (including the initial and final moment).

# Creative Fashion

Vanessa is a bit eccentric, and every time she goes outside, she wears two gloves of different colors.

In a box she keeps 23 pairs of gloves. Ten pairs are white, eight pairs are yellow, and five pairs are black. The lights in Vanessa's house are out, and she is completely in the dark.

How many gloves must she pick from the box to make sure that she can wear two of different colors?

# Hint

Try to find the worst-case scenario. How many gloves can Vanessa pick, at most, without having any pair of two different colors? Remember that she needs a right and a left glove.

## Answer

The worst-case scenario is when Vanessa picks all the right gloves (or the left ones), that is, she picks 10+8+5=23 gloves. After taking out one more glove, she will be sure to have the wanted pair. The answer is therefore 24.

# Halloween Costumes

It's Halloween and Chris, Brian, Michael, and Edward are out trick or treating. Their costumes make them completely unrecognizable. Here is what they say to their friend Jason, who is wondering who is hiding behind each costume.

Can you help Jason identify the four friends?

# Hint

You may find it useful to fill the following table with the information you have.

|        | Mummy | Ghost | Skeleton | Death |
|--------|-------|-------|----------|-------|
| Brian  |       |       |          |       |
| Chris  |       |       |          |       |
| Michael|       |       |          |       |
| Edward |       |       |          |       |

## Answer

Brian is the Death, Chris is the ghost, Edward is the mummy, and Michael is the skeleton.

# Part 2

# Intense Brain Workout

# Cursed Treasure

A crew of 20 pirates has just found on an island a treasure consisting of a large number of golden doubloons. At first, they divide the doubloons equally among themselves.

On the way back to the ship, four men are eaten by tigers. Eight more are killed during a tempest while they try to leave the island. The dead men's share of the treasure is redistributed equally among the survivors. That night one pirate gets drunk and falls into the sea bringing with him his share of doubloons.

Before they get to their destination, two more pirates perish after catching a fatal disease. The few surviving pirates then divide equally among themselves the doubloons of these two dead men. Now they have 350 doubloons each.

How many doubloons did each pirate have after the initial distribution of the treasure?

# Hint

How many pirates are still alive? How many doubloons did the drunk pirate bring with him to the bottom of the sea?

## Answer

After the first distribution, each pirate had 100 doubloons. There are only 5 surviving pirates, and thus a total of 350×5=1750 doubloons. Before the last two men died, that sum was divided among seven pirates, that is, each of them had 1750/7=250 doubloons. Therefore, 250 doubloons disappeared into the sea with the drunk pirate. It follows that the treasure consisted of 2000 doubloons. Each pirate had 2000/20=100 doubloons initially.

# Hospitable Island

After a storm destroyed their ship, three pirates become stranded on an almost-deserted island. Nobody lives on that island except Peanut the dog. Among the wreckage of their ship, the pirates have just found a box with some cookies. They decide to eat them the following morning for breakfast.

Being pirates, they know better than to trust each other. The first pirate gets up in the middle of the night and goes for the cookies. To keep Peanut quiet, the pirate gives him a cookie, then eats half of the content of the box. Later that night, the second pirate does the same thing: he gives one cookie to Peanut, then eats half of the cookies left in the box.

Finally, the third pirate goes through the same sequence of actions. In the morning, the pirates eat one cookie each and then toss the last cookie to Peanut. At that point, the box is empty.

**How many cookies were in the box when the pirates found it?**

# Hint

Reason backward. How many cookies were there before the third pirate ate his share?

## Answer

In the morning there are 4 cookies. Before the last pirate woke up there were 4×2+1= 9 cookies. Before the second pirate woke up there were 9×2+1=19 cookies. At the beginning there were 19×2+1=39 cookies.

# Four Towers

Jamal received for Christmas a box containing 100 wooden cubes of four different sizes. The edges of the cubes measure 3,4,5, and 6 inches, respectively.

One day Jamal, while playing with his mom, builds four towers, each one using only cubes of one size. The first tower only has 3-inch cubes, the second tower only 4-inch cubes, and so on.

Jamal is careful to make the four towers of the same height.

**How many cubes did Jamal use to build his four towers?**

# Hint

The height of the four towers will be a number which is divisible at the same time by 3,4,5, and 6.

## Answer

The height of the four towers is 60 inches. Blake used 60/3=20 3-inch cubes, 60/4=15 4-inch cubes, 60/5=12 5-inch cubes, and 60/6=10 6-inch cubes. In total, Blake used 20+15+12+10=57 cubes.

# Scouting Campaign

Misty has been instructed to conduct a three-hour scouting campaign. To move faster, she leaves her backpack behind with her group, which will be following her along the same path. Without the weight of the backpack, Misty moves at the constant speed of 6 miles per hour, while her platoon advances at 4 miles per hour.

After how long will she have to turn around to meet the rest of her company exactly three hours later?

# Hint

What distance will the company travel in three hours?

## Answer

Misty will have to turn around after 2.5 hours. In three hours, the platoon walks 12 miles. Misty can run those 12 miles in the first two hours. To make sure she is on that exact spot after one more hour, she should run forward for 30 minutes and then turn around.

# Underwater Mission

Grace, Leah, and Hunter, three experienced scuba divers, would like to explore an underwater cave. They calculate that it will take six hours from their diving point to swim across the cave and reemerge on the other side.

Unfortunately, they cannot bring enough oxygen to stay underwater for so long. Each of them can only carry one oxygen tank on their back. Additionally, they can bring a few more tanks secured to a rope. Leah can carry two tanks with the rope, while Grace and Hunter can each carry three. Every tank provides one hour of oxygen for one person. The explorers can drop any empty tank in the cave to be picked up in a subsequent expedition.

**How can they arrange the exploration to make sure that at least one of them reaches the other side of the cave?**

# Hint

Someone will have to go back to the base then dive into the cave again with additional oxygen tanks. Obviously, the three scuba divers will exchange tanks underwater, but keep in mind that Grace and Hunter can carry a total of at most four at any given time, while Leah can only carry at most three.

# Answer

Here is one possible solution. Let us say Grace is the one who will swim all the way across the cave. The three divers jump into the water together. After one hour, Leah gives one of her oxygen tanks to Grace, and then heads back. After one more hour Leah is at the base. She immediately grabs three new oxygen tanks and dives back into the cave. In the meantime, Grace has three tanks and Hunter two.

Hunter gives one of his two oxygen tanks to Grace who now has 4 tanks, enough to get to the end of the cave. Hunter turns around and starts heading back. He only has one hour of oxygen left, but on his way back he will meet Leah, who can supply him with additional oxygen to reach the base.

# Passionate Children

In a classroom there are 20 students. Four out of every five are in the sci-fi reading group, three out of five belong to the astronomy club, and seven out of ten are in the school chess team.

**What is the minimum number of students involved in all three activities?**

# Hint

First determine how many students are involved in each of the three activities. What does the sum of those three numbers represent?

## Answer

There are 16 students in the sci-fi reading group, 12 in the astronomy club, and 14 in the chess team. Adding up these three numbers, we obtain 42. In this sum, the students who are involved in 2 activities are counted twice, and those involved in all three are counted three times. Since 20×2=40, there must be at least two students who have been counted three times.

# Holiday Math

Joshua is spending some time with his family to celebrate Christmas 2021.

Over the holidays, he visits his aunts and uncles. Being fond of numbers, Joshua realizes that he and his 20 cousins were all born in different years.

The oldest, Sam, was born in the year 2000. Joshua also notices that the age of his cousin Maurice corresponds to the sum of the four digits of the year in which he was born.

**How old is Maurice?**

# Hint

If it were Christmas 2002, then Sam would be in the same situation in which Maurice is now. In fact, Sam would be 2 years old and the sum of the digits of his year of birth would also be 2, since 2+0+0+0=2.

## Answer

Maurice is 7 years old. In fact, he was born in 2014. Note that 2+0+1+4=7, which is exactly his age in the year 2021.

# Match Tricks

Ryan has laid out on the table nine matches in the way shown below. He challenges his sister Natalie.

"Can you make five triangles by moving three matches?"

Natalie thinks for a moment, then swiftly accomplishes the task

"It wasn't too difficult. The trick was understanding that not all the triangles have the same dimensions. Now let me ask you a question. Can you arrange the matches to make four triangles? This time all the triangles must be equal. Wait, you are only allowed to use six matches."

**Can you solve the riddles posed by Ryan and Natalie?**

# Hint

Natalie already gives a hint on how to solve Ryan's riddle. As for her question, you need to think... in three dimensions.

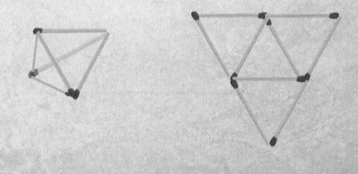

# Answer

Below are the configurations of matches that solve the two puzzles. Note that the second shape is a three-dimensional pyramid (more precisely, a tetrahedron).

# Chess Dilemma

Alexander, Caleb, and William took part in a chess tournament where every participant played against every other once.

Each of the three friends lost exactly two games. Every other player won exactly two games. In the entire tournament no game ended in a draw.

**How many people played in that tournament?**

# Hint

Start by considering the games that the three friends played against each other. How many other players do you need to add to make sure that each of the three friends lost two games?

# Answer

There were five participants to the chess tournament. In total, the three friends lost six games (two each). Since they played only three games against each other, more players are needed to account for those three extra losses. Remember that any additional player only won two games, and therefore one more player is not enough. Two more are needed.

A possible scenario with a total of five players is represented in the diagram below, where every arrow denotes a game. The head of an arrow corresponds to the winner of that game, while the tail to the loser.

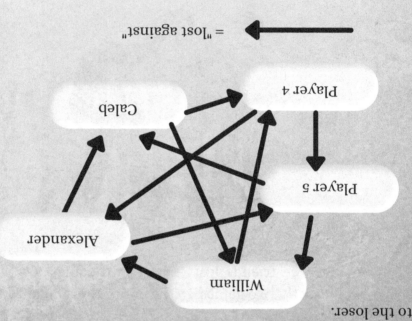

# Unreliable Musicians

In a group of ten musicians there are pianists, guitarists, and violinists. Each of the musicians plays only one instrument. Unfortunately, while some of them are honest, and always tell the truth, others are completely unreliable, and lie all the time.

An interviewer asks each of the ten musicians the following three questions: Are you a pianist? Are you a guitarist? Are you a violinist? He receives 7 'yes' to the first question, 5 to the second, and 6 to the third one.

**How many of the 10 musicians are honest?**

# Hint

How is the total number of 'yes' related to the number of honest musicians?

## Answer

The total number of 'yes' is 7+5+6=18. Each honest musician answers yes to one of the questions, while every dishonest musician answers yes to two of the questions. There are 2 honest musicians and 8 dishonest ones since 2×1+8×2=18.

# Coin Towers

Each quarter in Eugene's collection comes from a different US state. He hopes to achieve soon his goal, which is to have one quarter from each of the 50 states.

Today Eugene started playing with his coins. After arranging them in piles of seven, he noticed that he the number of coins he had left was equal to the number of piles. He tried arranging the coins in piles of nine, and again observed the same phenomenon: the number of quarters left was equal to the number of piles.

**How many quarters does Eugene need to finish his collection?**

# Hint

There aren't many possibilities to check: after arranging in piles of 7, the number of remaining coins can only be 0,1,2,3,4,5, or 6.

## Answer

Eugene has 40 quarters. In fact, 40 coins can be arranged in 5 piles of 7 coins, with 5 coins left, or in 4 piles of 9 coins, with 4 coins left. Therefore, Eugene needs 10 more coins to complete his collection.

# Bus Trip

Every hour a bus leaves the city of Annapolis heading to Bourgville. At the same time, a bus directed to Annapolis starts from Bourgville. The bus ride between the two cities takes 7 hours. Whenever a bus arrives in one of the two cities, it immediately turns around and begins its journey back.

Vivek is in Bourgville, starting his journey in a bus which arrived from Annapolis just a few minutes ago.

How many buses directed toward Bourgville will Vivek meet during his 7-hour trip to Annapolis?

# Hint

Vivek's bus left Annapolis 7 hours ago and will leave Annapolis again in 7 hours. How many buses depart from Annapolis between these two moments?

## Answer

Vivek will meet 13 buses going toward Bourgville: the one leaving Annapolis at the same time in which Vivek is leaving Bourgville, the six that left Annapolis in the six hours before Vivek's departure, and the six that will leave Annapolis in the six hours after Vivek's departure.

# Roman Identity

Morgan has just composed a mathematical identity in Roman numerals using a few matches. Her brother Jonathan enters the living room and stares at Morgan's work from the opposite side of the table. Here is what he sees:

Jonathan: "This identity is incorrect. Luckily, I can fix it by moving just one match."

Morgan: "But the identity is already correct!"

**How did Jonathan change the identity? Why is Morgan saying that the identity is correct?**

# Hint

Remember that Jonathan is approaching from the other side of the table.

Answer

# Superfluous Weight

Adrian has a big scale and a set of 15 different weights measuring respectively 1 lb, 2 lb, 3 lb, and so on until the heaviest one, which weighs 15 lb.

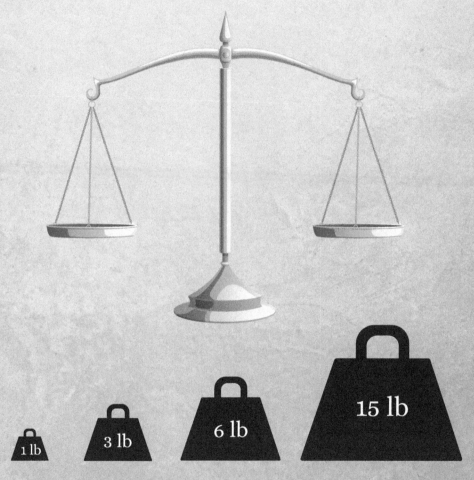

One day Adrian realizes that with just five of those weights he would be able to measure on his scale any weight in pounds starting from 1 lb up to 15 lb.

**What are these five weights?**

# Hint

The weights 1 lb and 2 lb must definitely be among the chosen ones. Which else?

# Answer

The five weights are: 1 lb, 2 lb, 3 lb, 7 lb, and 11 lb.

# Heavy Veggies

Alexandra is weighing vegetables. Here are two situations in which her scale is perfectly balanced.

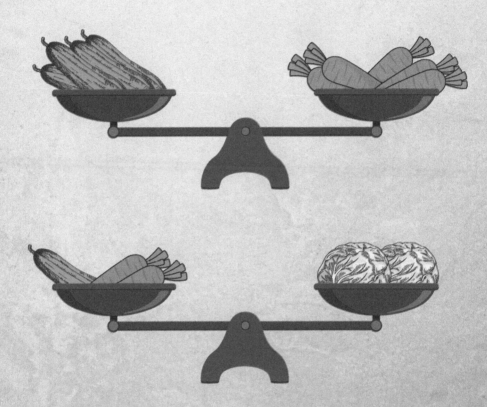

Alexandra now wants to weigh two heads of cabbage and three carrots against one carrot and four cucumbers.

**To which side will the scale tip?**

# Hint

You can replace two heads of cabbage with two carrots and one cucumber. You can also replace four cucumbers with five carrots.

## Answer

The problem reduces to weighing five carrots and a cucumber against six carrots. Removing five carrots on both sides, we obtain one cucumber against a carrot. From the first picture in the puzzle, we deduce that a cucumber is heavier than a carrot. Hence the left side is heavier.

# Geometric Webs

Steve the spider is inspecting his latest web, checking for possible defects. Last year, Steve spent the entire Winter hanging from the ceiling of a school classroom learning everything about geometry and math. Since then, he has been holding himself to the highest standards of perfection. He wants the length in inches of each segment of his webs to be an integer number. After concluding his inspection, Steve is satisfied: even this time, he was mathematically impeccable.

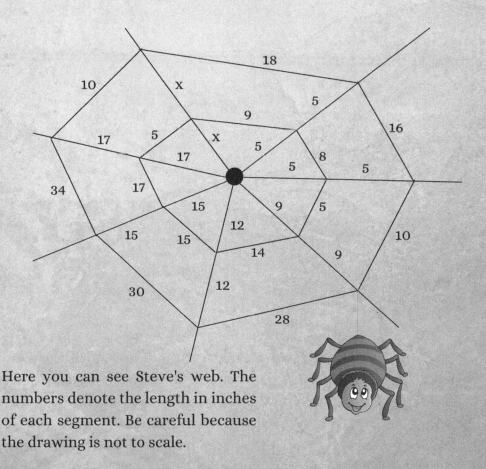

Here you can see Steve's web. The numbers denote the length in inches of each segment. Be careful because the drawing is not to scale.

**What is the length of the segment denoted by x?**

# Hint

Focus on the following piece of the web. It contains all the information you need to solve the puzzle.

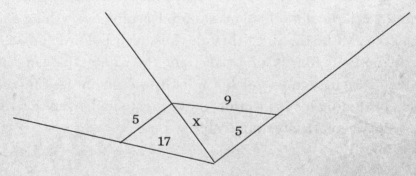

## Answer

In the section of the web highlighted in the hint we can see two triangles whose sides are respectively of length 17, 5, and 9.5, x. In a triangle, one side cannot be greater than the sum of the other two. Hence x needs to be at the same time greater than 12 and smaller than 14. Therefore, the length of x is 13 inches.

# Part 3

# Dangerous Territory

# Court Proceedings

Judge Morrison is facing a serious dilemma. The prosecution wants to call to the witness stand three members of a local gang. It is well known that in that gang there are only two types of people: those who always lie and those who always tell the truth. The prosecuting attorney has assured judge Morrison that at least one of the three is honest but doesn't know which one.

The judge has very little patience and wants to find out quickly whether the three witnesses are honest or not. He is aware that all the gang members know each other very well. He therefore summons the three into his chambers and asks, "How many liars are there among you three?".

After noting the answer given by each, the judge sends them out and starts pacing. He needs time to think. Having pondered over the situation for a while, the judge realizes that he doesn't have enough information to decide who is honest and who isn't. He returns to the witnesses and points at two of them. "You two, come back inside". He asks them "How many liars are among you two?". After hearing the answers, he knows exactly which of the three are lying.

**What answers did the judge get the first time he questioned the three witnesses? Who did he call in for the second questioning?**

# Hint

There is only one set of answers to the judge's first question that would leave him in doubt on who is honest and who isn't.

## Answer

The judge got the answers 1-1-2. These answers could correspond to two honest witnesses and one liar or to two liars and one honest.

The judge called back into his chambers one of the guys who answered 1 together with the witness who answered 2. Among these two, one is a liar, and one is honest. The one who will reply 1 to the judge's second question is the honest one.

# Birthday Party

Patrick is turning ten in a few days, and he has invited 9 good friends to celebrate with him. He stares at the list of guests and notices a peculiar fact.

The fist guest only knows one of the ten people who will be present at the birthday party (and that person, of course, is Patrick). The second guest knows two, the third guest three, and so on until the eighth guest, who knows eight. The ninth guest is Lauren.

How many of the people who will be at the party does Lauren know?

# Hint

You might find it useful to use the following diagram, where the first guest is represented by the number 1, the second by the number 2, and so on. A line between two people means that they know each other. The lines connecting the guests to Patrick have already been drawn.

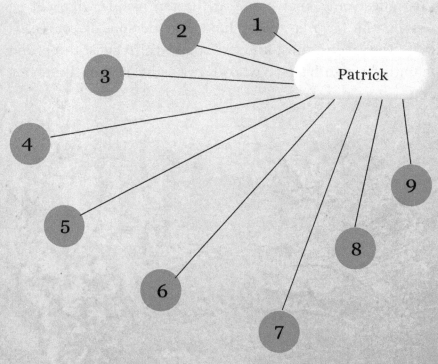

## Answer

Lauren knows 5 people: Patrick, plus four other guests. In Patrick's list, these four guests correspond to the numbers 5,6,7, and 8.

# Luxury Spaceships

Quke, who lives on the planet Gok, is looking into buying a new spaceship. The dealer is showing him around in the store. "We have three models. The basic one is the Dove. Then we have the Eagle, which costs double the Dove. The most exclusive is the Falcon, which costs double the Eagle. Production is limited. Currently we have in stock only half the number of Falcons as we have of Eagles, and half the number of Eagles as we have of Doves."

"How much would it be to buy all the spaceships you have here?" asks Quke. "Well, that would cost you a total 924 Quoins." For terrestrial reference, the Quoin is the official currency on Gok, and there are no fractions of a Quoin. "For your information, a single Quoin won't buy you anything here" adds the dealer.

**How much does a Falcon spaceship cost, at minimum?**

# Hint

How does the total price of the Doves compare to the total price of the Eagles? The Doves are half the price of the Eagles, but their number is double.

## Answer

If a product costs half of another product, but the number of items is double, then the total price of the two products is the same. For this reason, the 924 Quoins are equally distributed among the three models. Therefore, to buy all the spaceships of a certain model one needs 924/3=308 Quoins.

Note that 308=7x11x2x2. Hence, there are 44 Doves priced at 7 Quoins each, 22 Eagles at 14 Quoins each, and 11 Falcon costing 28 Quoins each. We could also have 28 Doves at 11 Quoins each, and so on, but we choose the first option since the problem asks for the minimum possible price.

# Self-Explanatory

Julie discovered a very special ten-digit number.

Its first digit tells you how many zeros appear in the number. The second digit corresponds to the number of ones, the third digit to the number of twos, and so on until the tenth digit, which tells you how many nines there are.

**What is Julie's special number?**

# Hint

Julie's number has many zeros.

## Answer

Julie's number is 6210001000.

# Math Magician

Yesterday at school, Malika told me she had developed incredible mathematical abilities and wanted to give me a demonstration.

She asked me for three numbers smaller than 10,000. I gave her 5482, 1450, and 2307. After writing them on the board, she asked our friend Lucas for two more numbers. He gave her 4517 and 8549. Malika had barely finished writing these two when she exclaimed "The sum of the five numbers is 22,305". I checked with the calculator, and it turned out she was right.

I was genuinely impressed, but also a bit suspicious. So, I asked her to repeat the game. This time, I gave Malika the numbers 1799, 2005, and 3580. When I heard Lucas saying 8200 and 7994, my suspicions were confirmed. Malika and Lucas had conspired to fool me, and I knew exactly how they were pulling it off.

**Can you understand how Lucas and Malika's trick works?**

# Hint

5482+4517=9999. Similarly, 1450+8549=9999.

## Answer

jLucas's numbers are based on the first two numbers chosen by the "victim" of the trick. To obtain his numbers, Lucas simply subtracts from 9 every single digit of the first two numbers given written on the board. For instance, if the first two numbers given by the victim are 5711 and 3168, Lucas says 4288 and 6831. The sum of these four numbers is 9999+9999=19,998. Thanks to Lucas's carefully chosen numbers, this sum will always be 19,998. To find the sum of all the five numbers on the board, Malika must look at the third number chosen by the victim and add 19,998, which is easy: just add 20,000 and then subtract 2.

# Dog Sitting

Mary agreed to take care of her neighbor Susan's five dogs while she is on holiday. Susan leaves on a Friday morning, planning to return on Tuesday night, after 12 days. Mary has the habit to give each of her dogs two cookies every afternoon. Of course, she will do the same with Susan's dogs.

For this purpose, Mary bought 192 cookies, exactly what is needed to give every dog (her own dogs and Susan's) two cookies on each of the twelve days. Unfortunately, Susan has to cut her holiday short, and she comes back on a Saturday night, after only 9 days. Mary keeps feeding the cookies to her dogs at the same rate of two per day.

**On which day of the week will Mary's dogs finish all the cookies?**

# Hint

Can you first figure out how many dogs Mary has?

## Answer

In 12 days, at the rate of two cookies per day, Susan's dogs would eat 5x2x12=120 cookies. The remaining 192-120=72 cookies are what 3 dogs would eat in 12 days at the rate of 2 cookies per day since 3x2x12=72. Therefore, Mary has 3 dogs.

During the first 9 days there are 8 dogs. They eat 8x2x9=144 cookies. There are 192-144=48 cookies left. Mary's 3 dogs eat a total of 6 cookies per day. It will take them 8 days to eat the 48 remaining cookies. Since they start on a Sunday, they will finish the cookies on the following Sunday.

# Picky Parrot

Peter is a pretty picky parrot, especially when it comes to food. He always eats 8 seeds, and then tosses the next 2 onto the floor. Then he eats 8 more seeds and tosses 2 more onto the floor. He keeps repeating this pattern until his food bowl is empty.

On Sunday, Adam cleans Peter's cage and fills his bowl with seeds. On Monday, Adam sweeps the floor, collecting all the seeds discarded by Peter. He puts these seeds back into Peter's bowl. On Tuesday, Adam repeats the same procedure. On Wednesday, there are no more seeds on the floor.

**How many seeds did Adam put into Peter's bowl on Sunday, at most?**

# Hint

Reason backward. How many seeds did Adam collect from the floor on Tuesday, at most?

## Answer

Adam put 8 seeds into Peter's bowl on Tuesday since no seed was on the floor on Wednesday. On Monday, the maximum possible number of seeds in the bowl (so that 8 would end up on the floor the following day) was 48.

On Sunday, the maximum possible number of seeds in the bowl was 248 (since 248=24x10+8).

# Asteroid Impact

The Earth is in danger! A massive asteroid is traveling toward our planet at the speed of 9 miles per second. Luckily, scientists and engineers at the EADD (Earth Asteroid Defense Department) have been working on a new defensive technology, giving us some chances of survival. They have built a special rocket ship that will be launched from the spot of foreseen impact of the asteroid. It will travel upward at an average speed of 3 miles per second and deliberately crash itself into the asteroid to divert its path.

Scientists wish it would be possible to hit the asteroid when it is far away from the planet. However, for maximum accuracy, they have decided that the impact with the rocket ship should occur only 6 miles from the Earth's surface, when the asteroid is about to enter the atmosphere.

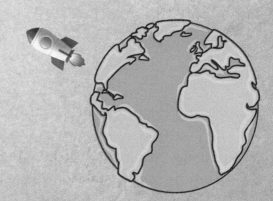

**How far away from the Earth will the asteroid be when the head of the EADD pushes the button to initiate the 10 seconds countdown for the launch?**

# Hint

How far from the Earth's surface will the asteroid be when the rocket ship takes off?

## Answer

It takes the rocket ship 2 seconds to travel 6 miles. In those 2 seconds, the asteroid travels 18 miles. Therefore, when the rocket takes off, the asteroid must be 6+18=24 miles from Earth. Ten seconds earlier, the asteroid is 24+90=114 miles from the surface of the Earth.

# Martian Mission

Rick and Miranda are the most daring astronauts of the 25th human mission to Mars. This time they have decided to climb Olympus Mons. Standing almost 70,000 feet above the surface of the planet, it is the highest and most impressive mountain on Mars. The astronauts start the mission with two backpacks of the same exact weight.

During the ascent, they stop to rest and admire the takeoff of a spaceship. While on this break, they eat all the food from Miranda's backpack, which as a result weighs one third less than Rick's. Rick then transfers some equipment to Miranda's backpack so that they are once again carrying the same weight. On their second break, they eat from Rick's backpack, which as a result is now three quarters the weight of Miranda's backpack, that is, 2 lb less.

**What was the weight of the two backpacks when Rick and Miranda started their ascent of Olympus Mons?**

# Hint

After the second break

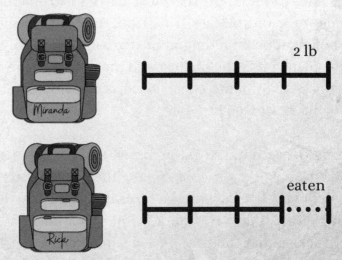

After the first break

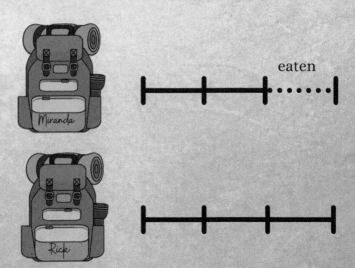

## Answer

Before the second break, they carried a total of 16 lb. In the first break, they consumed 16/5=3.2 lb of food. The initial weight of a backpack was 3.2 x 3=9.6 lb.

# Abstract Art

Liam shows his latest drawing to his friend Noah.

Look! I drew 17 squares.

Why do you say seventeen? Oh, I see. Sixteen plus the one in the middle. But wait, there are also four small ones in the center, and then the entire figure is also a square. And maybe there are others...

**How many squares can you count in Liam's drawing?**

# Hint

Be systematic. Go from the smallest size to the biggest.

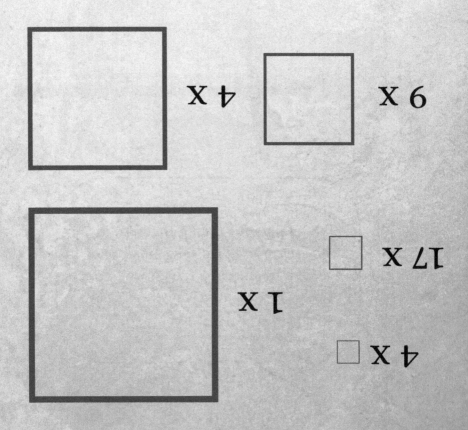

## Answer

There are 35 squares in Liam's drawing. Here is the breakdown by size:

4 x ☐
1 x ☐
17 x ☐
4 x ☐
9 x ☐

# Pick and Sum

Ethan is in front of a giant urn containing balls numbered from 13 to 2021, one ball for each number. He picks blindfolded two balls at the same time and then sums the two numbers on the balls.

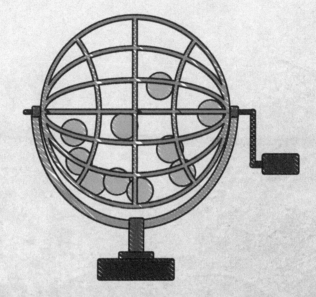

Here is an example.

**How many different sums can Ethan obtain?**

# Hint

What is the smallest possible sum that Ethan can obtain? What is the largest?

## Answer

The smallest sum is 13+14=27, while the largest one is 2020+2021=4041. All the numbers in between these two are clearly possible sums. How many numbers are there between 27 and 4041?

Well, it is like counting to 4041, but we are neglecting the first 26 numbers. Therefore, there are 4041-26=4015 numbers. Ethan can obtain 4015 different sums.

# Exceptional Memory

The game of Memory is played with a deck where each card appears twice. The deck is shuffled, and the cards are laid face down on the table. One move consists in flipping a card, then choosing a second card and flipping that one as well. If the two cards form a matching pair, they are removed from the game.

If not, they are returned to their original face-down position. The goal is to remove all the cards from the table in the least number of moves.

Zoe is exceptional at this game. She can always remember without failure the exact location of every card she turns. Now she has just laid down twelve cards to play a Memory game by herself.

**What is the maximum number of moves she will possibly need to finish the game (that is, to remove from the table all six pairs of cards)?**

# Hint

Think of the worst-case scenario, but at the same time remember that Zoe's memory is exceptional.

## Answer

The worst-case scenario is that in the first three moves no pair of matching cards appears. If this is the case, however, Zoe can start the next move by flipping one of the six cards she hasn't touched yet. It will form a pair with one of those she has turned before. Using her exceptional memory, Zoe will be able to remember where the twin card is. In this way, she will go on to pick up all six pairs one after the other. In total, it takes her 9 moves to finish the game in this worst-case scenario. In conclusion, she needs 9 moves, at maximum, to finish the game.

# Hungry Puppies

Isabella has four puppies named Diesel, Tank, Rocket, and Rambo. She is about to treat them with some delicious croquettes. Diesel, who is the most voracious, will receive 140 croquettes, while Tank, Rocket, and Rambo will get 120 croquettes each. Diesel, however, eats much faster than the others. He devours a croquette every second, while Tank, Rocket, and Rambo, each eat a croquette every two seconds.

When one puppy is done with his share, it will keep eating at the same speed but from the bowl of one of the others. This goes on until all the food is gone. The four dogs start tucking into their croquettes all at the same time.

**How many seconds will it take the puppies to finish all the food?**

# Hint

How many croquettes do the four puppies eat in a second? How many croquettes are there in total?

## Answer

There are in total 140+120+120+120=500 croquettes. Every second the puppies eat 1+0.5+0.5+0.5=2.5 croquettes. Therefore, it will take them 500/2.5=200 seconds to finish all the food.

# Fast Connections

In a distant planet there are four main cities. Each of them has recently been connected by magnetic elevated tracks to the other three.

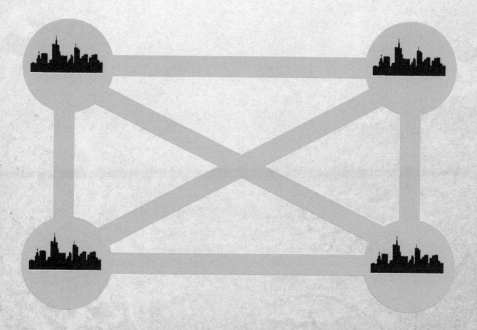

On these tracks three different types of latest generation trains can run: the LaserBlink, the ZeroGrav, and the TimeBender.

Zack, an expert engineer, is in charge of deciding for each route if it should be served by a LaserBlink, a ZeroGrav, or a TimeBender. He knows that the inhabitants of each of the four towns want to have their city reached by all three models.

**In how many ways can the four cities be connected?**

# Hint

Start assigning routes to LaserBlink, ZeroGrav, or TimeBender. How many can you choose before all the remaining choices are forced?

## Answer

There are six ways to assign the trains to the routes. Pick one of the four cities. Once you assign trains to two of the three routes starting from that city, all the other choices are forced. There are six ways to choose the first two trains (each model is denoted by its initial letter): LZ, LT, ZL, ZT, TL, TZ.

# Your Free Gift

We hope you enjoyed this collection of math brain teasers.

As a way of saying thank you for your purchase, we would like to offer you for FREE the ebook **12 Tricky Brain Teasers to enjoy with Family and Friends.**

Scan this QR code to receive your free gift now!

# Other Books in this Series

Make sure to check out on Amazon the other volumes of the series *Logic Games for Smart Kids and Teenagers*.

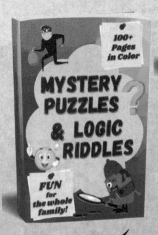

Made in the USA
Las Vegas, NV
30 April 2022